REBEL TALK

poems from the climate emergency
edited by Rip Bulkeley

with a foreword by Philip Gross

Extinction Rebellion Oxford 2021

Published by Extinction Rebellion Oxford November 2021

1 Poems © **The poets listed on the contents page**
2 Poetry in Times like These © **Philip Gross**
3 Design of illustrations and this edition © **BA (Hons) Graphic Design students at Oxford Brookes University**

Contributing students; Jemima Thomas, Philippa Lloyd, Jazmin Solomon-Siqueira, Ella Beales, Jessica Borrill, Dalal Elsamannoudi, Nazanin Azimi, Georgia Colson, Trizia Lee.

Sales

Information about how to buy *Rebel Talk*, including a fund-raising discount for multiple orders from campaigning groups, will be found on the inside back cover.

Information and media

Many of the poets are available for interviews or to read their contributions to this book. Media and other enquiries should be directed to the project team: **rebeltalk@protonmail.com** For information about climate-related performance events, featuring poets and other artists, please see your local and social media and especially local poetry groups.

In den finsteren Zeiten
wird da auch gesungen werden?
Da wird gesungen werden.
Von den finsteren Zeiten.

In the darkest times
will there really be any singing?
There certainly will be singing –
about those darkest times.

                                      Bertolt Brecht
                                      Svendborger Gedichte, 1939

Contents

8	Note to the Reader	Rip Bulkeley
10	Poetry in Times like These	Philip Gross

Earth

16	*Stag Beetles*	Steven Matthews
17	*Golden Child*	Annest Gwilym
18	*Still Life, Glasgow, 2019*	Alison Cohen
19	*Song of Praise*	Shanta Acharya
20	*Ocean Compost*	Rebecca Gethin

Loss

24	*Cassandra*	Kathleen McPhilemy
26	*Natural Neighbours*	Emma Lee
27	*Dead Ice: the End of a Glacier*	Kate Noakes
28	*California Screaming*	Julian Bishop
30	*Green Man*	Jules Whiting
32	*Great Northern Diver*	Caroline Jackson-Houlston
33	*drought*	Elisabeth Sofia Schlief
34	*Mountain Language*	Chrys Salt
36	*Wasps*	Jenny Donnison
37	*'Childhood wild-wood gems...'*	Kate Williams

Fear

40	*a revised draft*	Mandy Macdonald
41	*Global Warming Hits Margate*	Julie Lumsden
42	*Rainbows Behaving Badly*	Lesley Burt
43	*Gorge*	Deborah Harvey
44	*Starfish*	Rachel Burns
45	*In the Dry Season*	Gill McEvoy
46	*Inundation*	Elizabeth Rimmer
47	*Harbinger*	Paul McGrane
48	*After Air*	Martyn Halsall
50	*When the Rain Stopped*	Michael Bartholomew-Biggs
51	*Nature Study*	John Barnie
52	*digital digitalis*	Jordan Biddulph
53	*Winter Visions*	Paul Sohar
54	*Flood Rain*	Miki Byrne
55	*Intruder*	Rachael Clyne
56	*Let there Still be Birds*	Philip Dunkerley

Guilt

60	*Icebergs at Tate Modern*	Michael Shann
61	*Record Breakers*	Oz Hardwick
62	*Beachcomber*	Sarah J Bryson
63	*How to Bake a Planet*	Pete Mullineaux
64	*The Naming of Storms*	Jim Aitken
66	*Exchange with an Oak*	Susan Taylor
67	*Connected*	Jez Green
68	*When we get Old*	Amy Deakin
70	*For Planes to Fly*	Lesley Quayle
72	*Blind Man's Buff*	Merryn Williams
73	*Litter picking in the Mariana Trench*	Clare Marsh
74	*the small cabin*	Jack Houston
75	*This is the Last Time I will be a Bird*	Susannah Violette
76	*little funerals*	Ashley Chew
78	*How I Mistook the World for a Cow*	Cathy Dreyer

Struggle

82	*Youth Strike*	Aaron Lomas
83	*The Coral's Grief*	Janine Booth
84	*Why should we Care?*	Cheryl Moskowitz
86	*Glyphosate*	Sarah Watkinson
87	*'play-worn globe...'*	Helen Buckingham
88	*The Heat*	Jos Smith
89	*There are no Words on a Dead Planet*	Ken Evans
90	*this stanza...*	Nor Staines Shaw

Hope

94	*Spring 2018 – Rebel Dawn*	Joy Kenward
96	*Moving Images*	Leslie Tate
98	*'I saw our world upon the ground...'*	Valerie Leppard
99	*farigoule*	David Annwn
100	*Anthropocaust*	Rip Bulkeley
101	*'And what if all we do...'*	Justin Kenrick
102	*The Good Ancestor*	Daverick Leggett
104	*still life*	Emma Jones

106	Notes and Acknowledgements
108	About the Poets

Note to the Reader

Rebel Talk was conceived and created in Oxford by supporters of Extinction Rebellion. The call for poems was sent as far across the world as we could reach, and submissions were received from Australia, India, the United States, and several European countries as well as Britain. As the editor, Rip Bulkeley worked with Jane King and Madeleine Metcalfe in the project team, and with Wendy Davies and Phoebe Nicholson in a selection panel whose arduous 'blind' scrutiny of more than 300 submissions (with Rip speaking last and sometimes overruled) was essential for the book. On the design side, Neil Mabbs led a team of students at Oxford Brookes University who gave the book its powerful appearance by exploring some of the visual effects of waste-making processes, such as roller cleaning, during printing. Meanwhile several of the poets worked closely with the editor on minor aspects of their poems, so that we could all contribute as effectively as possible to the vital cause of rescuing ourselves from our own destructive folly. Everything was ready for publication shortly before the Covid pandemic threw the world into disarray. Even now, on the eve of the vital COP26 meeting in Glasgow, it is far from certain either that world leaders can act together to limit the now inevitable harm from climate change, or that climate campaigners yet know how to raise the storm of global public opinion that would make them do so. Many of these poems express their authors' sorrow and doubt in response to that situation.

Nevertheless, in their different ways the people involved with this book feel the need and the determination to go on trying, not least in the creative arts. *Rebel Talk* continues the cultural rebellion against governments, corporations, and other institutions which are still doing far too little about the climate crisis. We hope to set up or simply take part in performance events across Britain in which poets take part alongside performance artists in other genres. Please get in touch with us at *rebeltalk@protonmail.com* if you would like to take part yourself or suggest other performers, or better still, if you might be able to host such an event where you are. Please also keep an eye on our FaceBook page: *https://www.facebook.com/Climate-emergency-poems-Rebel-Talk-1054824368048880/* for announcements of such events.

In the book, the arrangement of poems into sections is very loose, because most of them contain elements of more than one such theme. Indeed the image placed before a section is just as valid a title as the word. An asterisk below an author's name means there is a note about the poem in Notes and Acknowledgements.

Lastly the editor extends his personal thanks to Philip Gross for his thought-provoking foreword, and to everyone else who helped create this book, including those whose poems could not be included. Will there be a sequel? Perhaps not from Oxford; but from somewhere else? You bet your life and the future lives of our own species and all others.

Rip Bulkeley
Oxford
November 2021

Poetry in Times like These

What is poetry for in times like these? See how naturally those words come to the lips: 'in times like these'... In their various ways, the poems in this book all seek to show just what, uniquely, these times are, and why it is once again so urgent that creative artists respond to the challenges they pose, in particular to the climate emergency.

'What is poetry for?' That's a fair question, always. From past history, we know that any simple, single answer turns out to be partial. Poetry can be, sometimes has been, almost anything. It is a channel for every kind of human speech, emotions – war cry or lament, comforter or agitator, love song or tribal chant of hatred, or sometimes a spacious calm in which we can sit and reflect. Its power is in the opposites it can contain – our need to be close up to life, and to hold it in perspective, equally. It is where we feel our thoughts and think about our feelings, where we can touch the abstract and watch the implications rippling out from sensual touch.

But 'times like these'... Just say those words, and feel the sense of rising urgency. If there is any gain in being the age I am, it's to realize that this is not the first time I have I heard myself saying those words. The first nightmare I remember comes from when I was 10. I looked out over my primary school playground and saw radioactive fallout drifting in like evil rain. This was 1960. The Americans and Russians were brandishing their threats across the world, and in a few years there came the Cuban missile crisis, with the world within an inch of nuclear war. I grew up listening for the three minute warning, which could be the end.

The feeling that we lived with, like a vault of crumbling buttresses above our heads, was dread – the same, perhaps, that people of the Middle Ages felt in times of plague. This is not to belittle either pandemic or nuclear terror (both those Horsemen of the Apocalypse are still alarmingly close at hand), still less the environmental dread which so many of us feel now, young people most acutely. Rather, let us believe that we all, whatever our age, have it in us to understand. Snap one thread of the complicated weave of things that support us, and the whole web comes unravelled. We have always known that, I suspect. If something has changed within my lifetime, it is our growing capacity to see that this warp and weft includes the widest, subtlest systems of the planet. If we have seemed to deny it, that's because it is too frightening, too paralysing, to let ourselves know that we know.

Each person has their way of expressing an urgent concern. It may be political debate, or scientific research; it might be action, might be art, might be both: some Extinction Rebellion demonstrations belong to a fine tradition of protest as performance art. Against all political odds, a handful of people can print an image on the public mind. The greening of London Bridge is in the same class of eloquent interventions as the struggles for women's suffrage and the anti-war protests of the last century. But in an age of rolling news and social media memes, are there still particular things that poetry can do?

It is not for me to say, of course. Poetry will shape itself around the needs of people who are drawn to use it. The interesting thing is that so many of us are, in many ways. You will see a range of these in this anthology, from passionate denunciation to an almost ritual grieving, from satire to a knowledgeable wonder at the world. None of these is out of place. Poetry that gives respect to detail, to the complexity of natural systems, is as much a contribution to the movement as a marching hymn.

'Times like these' need rallying cries. We need rhetoric powerful enough to call those in power to account. We also need a nurturing art, with quiet spaces in it, a taproot back into our values. Otherwise

all the passion and enthusiasm in the world will end in burnout
and disillusionment – that or the kind of zealotry that forgets how
to connect with most of the people around it, complicated and
compromised as we are. A truly ambitious movement wants not only
to win the rebellion but to build a culture beyond protest, one we
would want to live in, and would not need to be perfect to do so. We
can do better than yet another authoritarian utopia. We must.

But some tasks are too complex for politics alone. We need art. In
particular let me suggest we might need poetry. All the reasons why
I can't define what the poetry is for, those are the reasons why we
need it. One person's answer might be 'to express my feelings', or
'to tell my story'. Equally it might be 'to take me beyond myself... to
stretch my mind beyond my own experience'. One might be 'to play
with the glory of words'; another, 'it is the one place words have their
own silence in them'. Poetry can hold these contradictions, as we
must.

One key to this moment, for me, is that poetry is side by side with
science. You have only to pick up a copy of Oxford's splendid new
climate magazine, *Anthroposphere*, to see them in action together.
Both disciplines demand a dedication to seeing things in detail
as well as seeing them whole. There is no contradiction between
spiritual values and a vivid grasp of the materiality of life on earth...,
or if there seems to be, it is one that poetry can hold.

This book is not the last word. Why would it want to be? It is an
invitation to read, to pass on whatever moves you, to test your life
against these other voices, and to respond – in words, in action, in
contemplation or imagination. Poetry opens a door, but where that
door leads... that will be up to you.

Philip Gross
Penarth

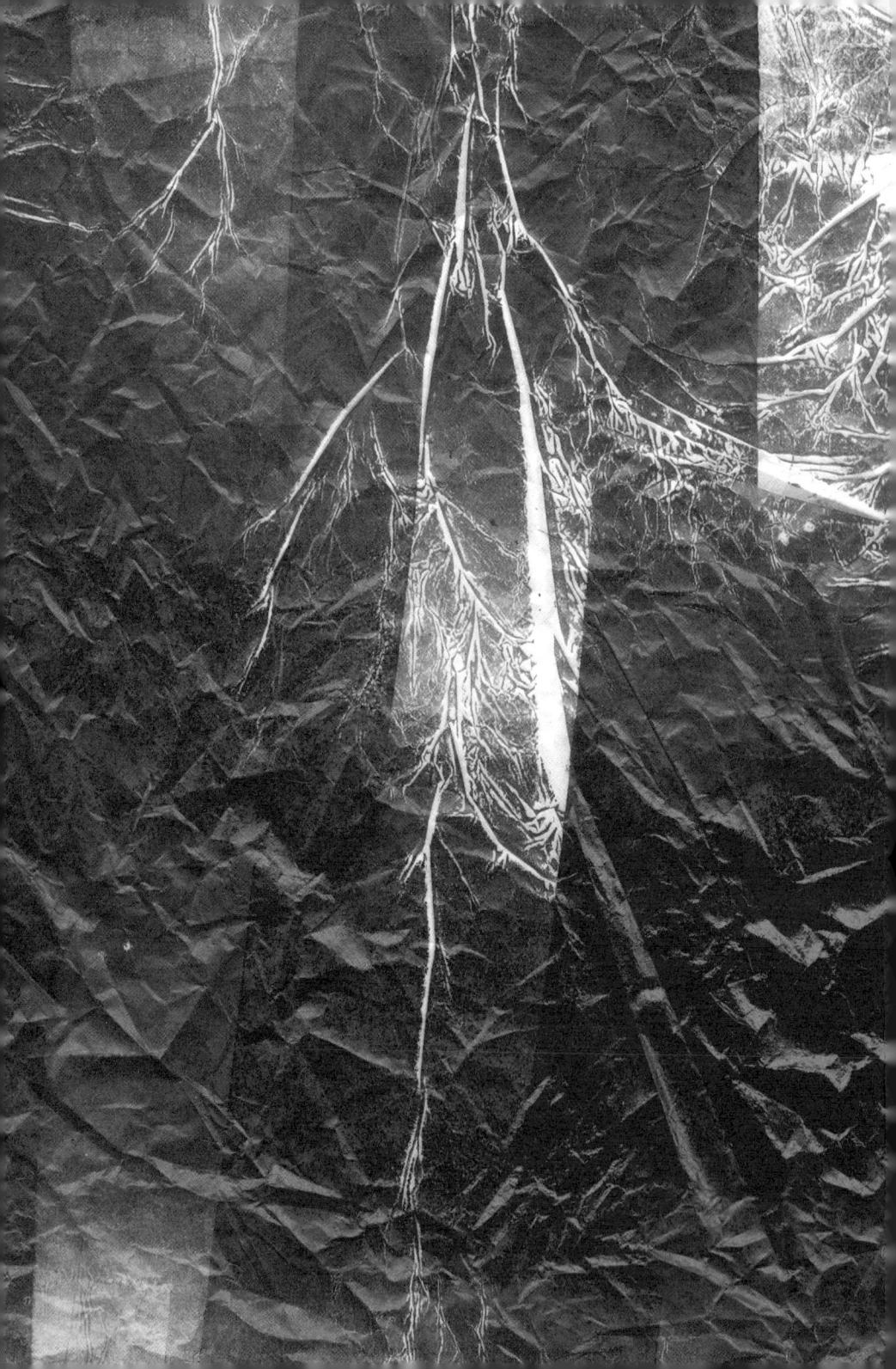

Stag Beetles

A memory: that July day
spent hunkered down, banking up earth
around the celery stalks, say
when I was ten; hating Dad's laugh

at the time it took me to do.
I wrestled the two steel buckets
hotly to the compost heap, through
his pea-canes and currant bush nets.

Where I was shocked. On the heap's top
two huge stag beetles had locked horns –
a black circling that did not stop
at a boy's shadow which had torn

across their world. I'm sometimes scared
in dreams by their strange aggression:
that clack of horns memories once heard
now just sad imagination.

Steven Matthews
Oxford

Golden Child
Raja undulata

S
omewhere,
right now, a spinulose
angel flutters on the sea bed
in her camo dress, swirls and spots
applied in pointillism. Sometimes spotlit
by a wobbly sun, from snout to wing tip she
undulates gently like a handkerchief blowing in
the breeze. Powered by her rudder tail, she burrows
and melts into soft blonde sand. Beauty queen of rays,
she hides her cartoon face underneath where she
grins with 50 teeth. She bears children in a purse
fit for a mermaid, miniatures that need
no mothering. Golden child, I pray
you don't go the way of
the golden
toad,
no
oo
oo
oo
oo
oo
oo
oo
oo
oo
oo

Annest Gwilym
NW Wales
*

Still life, Glasgow, 2019

A heaviness of
lawn with laundry hanging
limp along a washing line, shrouded
by a hundred years
of tenements and trees.
Sixty feet above,
shading rhubarb and a guelder rose,
the gravity of sun is caught,
snagged up in a canopy
of bark and broken green,
painting pock-marked shadows
on the grass.

An ash is leaning on the garden wall,
letting all her hair down:
a woman lost in grief, pausing,
singing blues,
branches swathed in moss,
spelling through her roots
and stressing to mycelia
the worry of her pinnate leaves,
trembling and resilient
in the passive city's air.

Her crown has alopecia,
looks much lighter
in these summer months,
bears a stillness
of too many leafless upper bones:
silhouetted clothes props
pointing at the puckered sheets of sky.

Alison Cohen
Glasgow

Song of Praise
from prayers in different traditions

Praise the stars in their constellations
for knowing their place, yet blessing all migrations.

Praise the sun, powerful yet unwavering
in its journey across the sky, light pulsing
through clouds, mists – life sustaining.

Praise the moon always true, waxing waning,
constant in its daily transformation.

Praise the earth as it moves on its axis –
inner and outer cores holding on to each other,
partners on the dance floor, steady as they go.

Praise day and night, mere limits of our perception,
and death, a release from our earth-bound vision.

Praise the sky, air, ether; praise the universe
for awakening us to worlds beyond our imagination.

Praise water in all its forms, giving and taking –
blood flowing through continents of bodies.

Praise plants sun-facing, light-changing,
breathing in carbon, green deities in meditation,
giving us oxygen, expecting nothing in return.

Praise the eye of the guest – clear, observant.
Praise the giver of life – almighty, benevolent.

Praise every species in our planet
masterpieces of evolution –
rich, rare, wild, keepers of infinite secrets.

Shanta Acharya
Highgate

Ocean Compost

Not what I expected —
it's gentle and fluffy
flocculent
whale crap rises from arses
pluming to the surface,

where it floats and disperses in a pink-red smoke
among ribbons and fans of phytoplankton
smaller than the width of a hair:
wanderers, daylight-feeders, sea jewels,
silica-slicked diatoms, drifters

that revel and bloom
in the iron-red riches of cetacean scat,
using up carbon dioxide, breathing out oxygen.
Adrift in a galaxy of other microscopic creatures
their atoms jiggling and throbbing as they circle one another:

bristly arrow-worms, shrimp-like krill,
paddle-worms that suck out juices,
sea gooseberries with retractable tentacles,
sea-angels or slugs catching prey with one foot,
sea-butterflies dragging a fishing net of mucus,

along with ones who swim deeper,
and all of them feeding fish, basking sharks and baleen whales who,

on their long migrations
from north to south and back again
with their calves,
suck in this gravid water, then crap
as they sing to one another.

They deep-dive into the dark, mixing the gases
of waters as they rise up to breathe,
mending the world's air
helping create clouds
just by being what they are.

Rebecca Gethin
South Dartmoor
*

Loss

Cassandra

The woman is slim, young, soldier straight;
she strains her shoulders backwards
against the straps of the baby sling
though the baby is still so new
its arms and legs barely protrude;
the man's spine curves the other way
his burden the rucksack of baby necessities
nappies, wipes, a bottle, the just-in-case
full set of clothes; their companion
another woman carries nothing and her long
curly hair lifts behind her.

Such a beautiful day they have come to the rec
the playground they are still not quite entitled to;
they sit on a bench to feel sunlight on skin
escape from claustrophobic hours in the dark
sleeplessness, the insatiable recurrences
their child has introduced them to;
relaxed, reassured they can taunt their friend
with the tiny perfections of fingers and toes
the yearning tenderness evoked by any small mammal
challenge her choice, her earth-saving earnestness
decision not to have children.

Such a beautiful spring they hardly notice
cuckoos are absent, bats do not fly
swifts arc and wheel in reducing numbers.
She sits beside them, marking all this
files it along with other horrors
she knows they'd rather she didn't talk about;
she lets her hair droop round her face
like tattered curtains to the world's future
while across the park busy house martins
come and go from the eaves of the newbuilds
behind the wall, beside the lane to the meadow.

                                            Kathleen McPhilemy
                                            Oxford

Natural Neighbours
Belushya Guba, Russia

There are new unofficial rules;
bears move faster than legislation.

Don't go out alone. Only move in cars.
Bears are forming street gangs.

Be prepared to fire flares or stun grenades.
Bears are becoming bolder.

Reinforce doors, check stairwells
before the bears beat you to it.

Lock away refuse or burn it.
Bears forage when sea ice melts.

Don't run, even when instincts scream.
The bears will think you prey.

Don't feed them. They will return and bears
will rely on handouts instead of hunting.

We are getting used to being neighbours.
At least white fur makes them visible;

in other areas, the tigers are moving in.

Emma Lee
Leicester

Dead Ice: the End of a Glacier

Winter's snows are its inhalation; a freezing
draw, replenishment.

Water flows in summer as if from everywhere;
a thaw rich with minerals.

Moraine grinds bedrock in slow descent
and retreat; a seasonal breathing.

It has been like this for twelve thousand years.

A new field is strewn with erratics; rocks
perched on castles of sand.

Long summers now take more of the blueness
than late winters give.

And you, my beautiful daughters, will not
marvel with your children at its depth,

will not hear the echoing melt-song, or
be shocked by coldness under your palms.

Storehouse of all our seasons, the ice is dying.
No, my darlings. It is already dead.

Kate Noakes
London

California Screaming

All the leaves are brown and the sky is grey:

another papery day breaks, our hands torn wires
severed from sunlight
and only old songs fill the smoky void —

I pretend to pray

for a rain that never came. Now everything is ashes

get down on my knees

knowing I was to blame
only to be silenced by flames. Bodies on fire
but I was the survivor, I heard our everyday songs become screams.

I went for a walk on a winter's day

turned unwittingly into a burnt-out zone,
the place we called home, a town called Paradise
swept away by wildfires of our own making.
It isn't a dream — I was in California

and I was screaming

and I was screaming:

it isn't a dream — I was in California
swept away by wildfires of our own making,
the place we called home, a town called Paradise
turned unwittingly into a burnt-out zone —

I went for a walk on a winter's day

but I was the survivor, heard our everyday songs become screams
only to be silenced by flames. Bodies on fire
knowing I was to blame

get down on my knees

pretending to pray for a rain that never came. Now everything is ashes.

I pretend to pray

and only old songs fill the smoky void –
severed from sunlight
another papery day breaks, our hands torn wires:

all the leaves are brown and the sky is grey.

Julian Bishop
Hertfordshire
*

Green Man

It started as a remembrance,
lying on the lawn,
always in the early morning.

Bone cold, he'd watch
how his clothes
would suck up any wetness;

bloom his shirt
from camouflage to meadow.
He loves the scent

of damp greenness,
sinking into the ground,
becoming part of its tangled roots,

squeezing down his breath,
his heartbeat, till he's no more
than another blade of grass.

Sometimes, because
he has the choice,
he can stop being green.

He gulps air, holds it
in his lungs till his chest fills
with world,

then exhaling, bit by bit,
his entire body rises up,
yearning.

Of course everyone knows
he's crazy. The earth hasn't grown
grass for twenty years.

Jules Whiting
Didcot

Great Northern Diver
'Least concern but declining'

Listen: the loons are sirens, calling you
to a land of long light on dark waters –
a moan of loneliness and of desire.
Drown yourself down to the roots of lilies
where gold light filters through the peaty pools.

See black, see white, think Fred Astaire's screen test:
'Can't act, can't sing … can dance a little.'
                                                                 Then
recollect the Great Auk, fellow bubble-
trailer, deep-sea-flier, fish-follower.

Back low in the water like the dark side
of a wave, spangle-necklaced, ruby-eyed,
but crippled on man's element, the earth.
Each warmer year brings more to hobble from –
the holiday dogs, the raiding raccoons,
the breath of fiery air and forest ash.

Listen: the last loon is a siren,
                      vanishing aukwardly.

Caroline Jackson-Houlston
Oxford

drought

over the fields
dust swirls
sand devils dance
and spin
round and around

on the horizon
a red sun
puts glow-fingers
over the vastness
denying life to
the evening breeze

on the edge of the pasture
not far from the herd
dark feathered figures
wait motionlessly
for the meal
they will be sure to get

Elisabeth Sofia Schlief
Bonn

Mountain Language

So used were they to broken things,
smashed oil lamps, wine jars,
rattled urns, the creak of houses,
tumbling masonry,
they just shrugged off the portents
of a mountain's rage,
repaired their fallen walls and soldiered on.

Those who could read the writing in spring wind,
sun-scribbled messages on ice
told the prospectors it was dangerous,
warned of a snowpack in the mountain's throat
before it raised its mighty voice and spoke.

So when Vesuvius spewed its guts
or, fast forward two millennia,
when a slice of snow
broke loose from nature's moorings
smothering everything,
it was too late to call for gods,
too sudden for each moment's fending off.

Some were found curled to foetuses,
or fused together in a last embrace.
Some frozen in a running shape,
or curled like cats asleep
heads on their forearms under ash or snow;
others tossed like puppets down the slopes,
hands raised, bones broken, limbs agley,
perplexed perhaps to hear their mountain roar
before the snow or lava outran everyone.

'Do we learn anything from history,
the stranded polar bear, the rising sea?'

When writing this I found a photograph,
an unnamed man in glasses,
balding, elderly.

He holds against a sky blue shirt,
a plaster cast –
the body of a child from Herculaneum,
hands splayed across the tiny back,
so tenderly
you'd think the child and his heart might break.

Chrys Salt
London / SW Scotland

Wasps

Last year wasps were absent from picnics
from garden meals in summer drowse,

did not gatecrash in their tiger garb,
nor stare at us with ink dark eyes.

They didn't trouble us
with sting and buzz,

crawl the rims of floral plates,
sip sweet wine from the brimming glass.

No weaving flights or stumbling gaits,
drunk on gluts of fallen plums.

Jenny Donnison
Sheffield

Childhood wild-wood gems –
hums, glints, scents, hints – are slipping
from our youngsters' grip.

Kate Williams
Cardiff

FEAR

a revised draft

after millennia
and millennia of watching
and wondering
where she'd gone wrong
Gaia
went back to the idea she'd had
at the very beginning

and when she spotted the first
inquisitive simian
in its sun-shot edenic jungle
painfully curving back its spine
in the wrong direction
forcing its pivoting hip-joints
to bear an unnatural load
just to get at a particularly
rosyjuicy fruit without
having to climb

she rewrote the plan
stirred the gene cauldron
stamped them all out

Mandy Macdonald
Aberdeen

Global Warming Hits Margate

FUNLAND, Dreamland, Contemporary Art Gallery,
wartime bunker – *Lisa loves Dave*, fading with him
inside a thin green heart next to a swastika and *shit
Tracey is a slag* – all waiting for the triumphant sea.

Julie Lumsden
Chesterfield

Rainbows Behaving Badly

Slumps, suddenly red, throws its weight
across two farms and a village, puffs
gunmetal breath across sky, straddles
poplars ranked around a field's perimeter;

compresses spire and weathervane, crushes
chimneys and garden sheds to liquid green,
surges down sloped streets, collapses
the bridge's arcade into indigo chunks.

On the south bank, a farmer shakes a fist
at the torrent, hauls sheep to higher ground;
the priest, white-faced on the north side,
pulls on waders, calls on his flock for calm.

Lesley Burt
Christchurch (UK)
*

Gorge

Whitebeams and rock cress, speedwell and squill
think they've grown here for ever.

Only the stones remember
the webbing of sunlight and water
sharks silhouetted against the sky

fall open like books to share their story
show their collections of teeth and shells,
jostling corals and crinoids pressed like wild flowers.

They're trying to warn us. The smell of salt is in the air,
coffined creeks and culverts stir
under tarmac
                        black water rising

    this shiftless city lulled on dreams
the mud-bound boat under Stephen's church

                              starting to drift

Deborah Harvey
Bristol
*

Starfish

The children are frightened, thousands of starfish litter the beach.
It has been snowing all week. Frozen pipes have burst.
The snow is melting. Cities are slowly filling with water.
For the first time in twenty-five years, a polar bear cub is born in a zoo.
In the Arctic, the sea ice is melting. Polar bears are drowning.
The children follow the trail of dead sea creatures. We take photographs
of the children holding the dead starfish in their hands.

Rachel Burns
Durham City

In the Dry Season

The earth is parched and shrinking,
the grass has given up its hold.

I think about water – we are all thinking about water –
there is a hosepipe ban in force, and stringent
warnings to be frugal in our use of it.

I measure out mean cupfuls from the tap,
imagine (my body ninety percent this element)
each cup as hand, foot, forearm.

On the windowsill above the sink
a jade plant, fat green money tree,
is flourishing,

every leaf a reservoir of hoarded wealth,

while *rain* now seems to me a word so beautiful
I roll it on my tongue like a wildly expensive taste,
chant it like a mantra:

rain, rain, rain, rain,

as if in calling I could make it come.

Gill McEvoy
Yealmpton, Devon

Inundation

There has been a flood. The rising blue
fills the hollow space between the trees,
and washes over hillocks with a strange
still completeness, as if the sea had learned
to flow uphill. I will remember this
when pods of dolphins swim through Marble Arch,
and divers pick up rusted mobile phones
from silent oyster farms on Princes Street.
I will remember then the sunlit spring,
and Aberfoyle's green braes, drowned in bluebells.

Elizabeth Rimmer
Stirling

Harbinger

Walthamstow was welcoming a snowdrop
   though we hadn't even seen a sign of winter.
Daffodils too, bees and budding branches,
   hedges fat with music, soft with midges.

No one talked of snow or bet on Christmas being white.
   We'd forgotten what we looked like in a coat,
warm as butter when we should be buttoned down.

Someone should have asked what the fuck was going on,
   marching with a bedsheet for a banner;
but now the gates were off their hinges,
   spring banned the random use of water and the hawthorn flower,
summer brought fruit fall, abandoned beaches,
   storms with names that came crashing through the alphabet,
first witness to a swallow heading south.

Paul McGrane
Walthamstow
*

After Air

Edging from forest he found cleared sky grown endless,
ground lined with arrows, fading like peeling snow,
stretching to distance; white light like an indictment.

No wolves, or evidence of gun-metal strangers
who might snatch his pack. He followed miles of fencing,
sagging and barbed, marking out separation.

Gap to duck under; ruins in hazy distance;
no singing in this stacked, late English sky.
Striding, he was out in space untrodden since The Downfall.

Sunlight had squared off shadows when he arrived,
his boots unsteady over buckled brickwork.
He cleared collapsed signs, buried by hail-storm glass,

found word in shadow, *Airport*; other symbols,
somehow sheltered from demolishing weather,
read *Long-haul, Aliens, Duty-free, Departures*.

These he could trace from the book by his water bottle.
He decided to pitch for the night; a ghost wind,
banshee along passageways, driving him into a dream

of thousands towing their lives, wheeled towards wings,
answering adverts selling change of weather,
writing obituaries in vapour trails.

Next day no sound from parched sky, aching woodland.
He kept the sun behind him, tracking north
seeking some land to plant in warming winter.

Nothing in skies now larger than a circling vulture;
nothing to guide him after roads had crumbled;
nothing through forests, except danger of strangers.

No

Martyn Halsall
Santon Bridge, West Cumbria

When the Rain Stopped

When the rain stopped
nothing much was said for quite a while.

God's rainbow undertaking
not to flood the earth again
made no one spring to the conclusion
a drought might drag it to an end.

But as the autumn came
and went without a downpour
rationing requests
acquired a non-apologetic edge;

then we understood
we'd less and less of clearance over
once well-covered reefs.

It loomed obvious to all of us
that being in one boat
guarantees no single passenger –
let alone the vessel –
safe arrival.

Michael Bartholomew-Biggs
Islington

Nature Study

Picture the mountain beetle rolling sheep's dung,
a tiny Hercules, black as determination;

picture the hovering kestrel and the feast it makes,
extruding chitin on a fence post on the Rholben;

picture ourselves waiting for the curtain-call,
life the amateur drama group's poor little show,

wondering whether, now the play is over,
anyone will cheer as the earth stacks the chairs.

                                          John Barnie
                                        Aberystwyth

digital digitalis

digital digitalis
pixelated nature
void of life
binary DNA
electrical leaves
cable stems

screen captured
.JPEG
image saved
edit/photoshopped
cropped
no background
colourless filtered

image realistic
touch screen
LED indent
warm feel
rough perspex
behind monitor
hot hard plastic
hold down button
power off

Jordan Biddulph
Hertfordshire

Winter Visions

This winter brings unquiet dreams,
yet we can't help dreaming on
even while the bears remain awake;

just today, the fifth of January, I saw
a mama bear with her two cubs
browsing right beside Millbrook Road
in Worthington State Park, New Jersey.

I never talk to bears but today I should've
stopped and asked them
what was keeping them awake,
what dreams had slashed the cobwebs of their sleep.

Maybe they know something
we too should know;
the visions of insomniacs are etched by lightning
into real clouds in a fragile sky.

Or else these three bears I saw were somnambulists
like us who want to keep on dreaming
that this winter is as real as any other,
no better, no worse.

Paul Sohar
Warren, NJ

Flood Rain

For thirty-six hours rain
threw fine spears at the ground.
Each exploded back.
Danced centimetres high.
Each hour saw leaves
ripped away
like post-protest banners.
Branches bared
to their bones, blackened
by soaking and each step taken
squelches, even on pavement.
Drains do their best
to guzzle excess and lie with litter
draped over their iron grilles.
Backing and packing
into gutter-dams that cause
small lakes to lie
like silver plate upon tarmac.
Locals call it 'flood rain'.
It's thick, heavy, powerful.
Comes with drumming thunder
and a lightshow jagged
as barbed wire.

Miki Byrne
Tewkesbury

Intruder

Doe finds her way into my garden, wreaks havoc,
after Wilf clears the brambles by the hedge.
We stare at each other, consider our moves.

She's stripped tips of rosebush and apple.
Gently, I encourage her to leave, mark the spot
where she squeezes through gaps.

Next night, in pyjamas and wellies,
I roll out chicken wire in a rush,
chase her off with a handclap.

Doe persists. Higher fencing is needed.
Meanwhile the ash tree, next door, is dying,
leafless tips jut, like unanswered prayers.

Fences won't stop the starving;
all will migrate to higher ground
until the soil itself abandons us.

We keep our territories for now,
but our eviction has been served.

Rachael Clyne
Glastonbury

Let there Still be Birds

After the holocaust,
after we have done away with ourselves,
let birds survive. They require so little of the earth,
just a small place to build a home,
and food, grubbed, picked or pecked
from anywhere.

I like to imagine their flight across the sky
after we have gone,
flashes of colour, and their songs
still sounding across the bruised planet
empty of people.   No one left
to see, or hear.

It's a thought worth clinging to.
After the holocaust, let birds survive.
The earth deserves birds.

Phil Dunkerley
Bourne, Lincs.

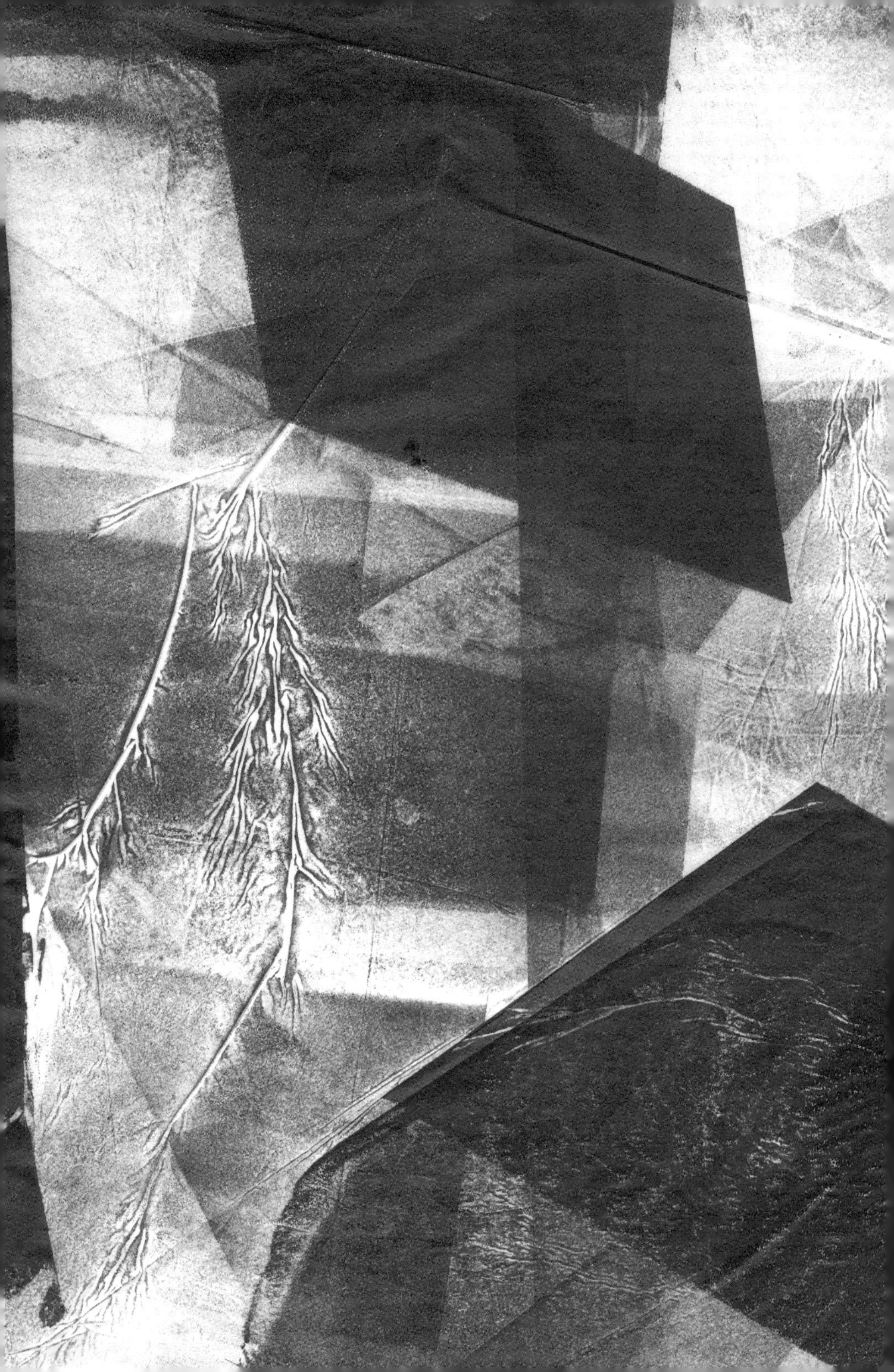

GUILT

Icebergs at Tate Modern

All the pubs are plumped and shrill with Christmas,
so we saunter down the South Bank under

the plane trees' dripping light, then see their blue-
tinged luminescence lovely in the darkness.

Moving closer, we're shamed by their shrinking,
animate forms. Like young boys embarrassed

into giggles by something serious,
we see who can press the cold the longest.

Michael Shann
Walthamstow
*

Record Breakers

I remember Roy Castle grinning on teatime TV:
the longest sword swallowed (23 inches),
the world's tallest woman (7 feet 7 inches),
the world's strongest beer (67.5%),
Ross and Norris McWhirter, stopwatches at the ready,
the world's fastest mile (3:43.13),
the most drumbeats in a minute (1,208),
the most raw eggs cracked and eaten in 30 seconds (9).
A pandemonium of children. You needed dedication,
dedication, to be the best, to beat the rest.

But now the whole world joins in without trying,
and teatime bursts with breaking records,
with fear-faced newsreaders and meteorologists
reeling off highs and lows: the warmest year (2016),
the fastest melting glacier (57 cubic miles per year),
the highest atmospheric CO^2 concentration (2019),
the fastest species extinction rate (1 every five minutes).
A pandemonium of shuffled papers and averted eyes.
Wait: where's the catchy theme music? The cheering?
Where's the dedication, dedication? The stopwatch
stops.

Oz Hardwick
York

Beachcomber

At low tide I meander from towels to water's edge
from shingle to sand, eyes down, stop now and then
to pick up pebbles, sculpted driftwood, fine shells.
Each piece is considered — some discarded

others held for their warmth from the sun, their mass
felt in the palm of my hand. As I walk I fill my pockets
with small pieces of sea glass, let my mind work
on mosaics of frosted greens and browns.

I carry with me a bag for objects of a different sort —
plastic beer mugs, forks and spoons, sandwich wrappers,
empty cans of cider, coke, beer, metal caps,
crisp bags, fag ends. Blue nylon string.

Today's haul includes the remains of a picnic
half-buried in sand — scrunched foil, the rusted rack
from a one-time barbecue, bottles, biscuit packets,
and a knot of bright yellow rope as big as my fist.

I walk back to the towels. On my way I find
an old house brick, smoothed by the waves,
a bleached cuttle-fish shell, and a heart shaped
air bladder among the tide-line bladder wrack.

I crouch down — it's still damp and smells of the sea.
I take it between finger and thumb to press it, then release,
press and release. I feel the tension, sense its robust fragility.

Sarah J Bryson
Kirtlington

How to Bake a Planet

First warm the oven – dust
and prepare a suitable work surface.
Sauté all ingredients in lava, allowing
the mixture to thicken and harden
into a crust of pasta-like sheets.
Use rainwater to make stock,
adding seaweed, kale – boil
and cool repeatedly. Simmer.
Meanwhile, gather fresh greens,
bone the fish, marinate the meat in
juices from sun-ripened fruits, whole grains,
nuts, garlic to taste – pour liquid over
dish until it almost drowns.
Increase heat gradually, stirring continuously.
Flambé the mushrooms!
Throw everything into a sealed container
with generous lashings of crude oil.
Set timer –
bake indefinitely ...
(Drink a glass of wine, watch a movie, disaster ones are best.)
Remove contents,
add a final pinch of salt –
serve chilled.

Pete Mullineaux
Galway

The Naming of Storms
after Henry Reed

Today we have naming of storms. Yesterday
we had heavy rain. And tomorrow morning
we shall have what to do after heavy rain. But today,
today we have the naming of storms. Sodden leaves
float like brown paper in all the neighbours' gardens.
And today we have naming of storms.

This is low pressure here. And now this
is high pressure there. Check out the clear
difference in packed and less packed isobars
on this chart. The branches float down
the burst river along with cars and caravans
and the remains of some neighbours' houses.

This is the weather warning chart. We are at
amber today which means we must be prepared.
Red would mean take action due to excessive
high rainfall. Start swimming, if necessary. Out at
sea oil tankers continue to shuffle their heavy loads
across the oceans to keep our cars running efficiently.

And this is the chart with all the names of all
the storms chosen by you. We started off with Abigail,
then Barney came, then Clodagh, followed by Desmond.
Eva was wild and so was Frank. What on earth will Nigel do?
Some places have had no rainfall at all and we seem unable
to harness our excess rain to make their deserts bloom.

This is where we believe lie the remains of Noah's ark,
near the summit somewhere of Mount Ararat in the Caucasus.
What with our failure to support renewable energy and our
palm oil plantations replacing jungles and our forests on fire,
IKEA has just brought out a new range in family arks. And this
now ends today's naming of storms. Enjoy the rest of your day.

>   Jim Aitken
>   Edinburgh
>   *

Exchange with an Oak

I didn't plan how to feel
when leaving this student bedsit
on the first floor, up close
to an oak's sturdy limbs.

The tree has, just lately,
unfurled all its aerofoils;
I would have thought
enough to fly.

If it could perceive me better;
an earthling, therefore part
of the race with many tricks
to bring it down, it would:

change the colour
of its flag to red overnight,
unroll a multilingual 'enough'
from its leafy tongues,

remove the saving grace
of its oxygen from my air.
Until the last minute, I go on
drawing gift after gift

from this oak; an old routine
we continue for every day
I am here. Transpire, partner,
estrange, reconfigure.

Susan Taylor
Dartmoor

Connected

In the moment
each gesture corresponds
to the spreading of the universe,
the dancing of solar flares,
the turning of the earth,
the diving of the manatee,
the whirling of rotifers,
the buzzing of electrons.
How deep does this go,
this rabbit hole of oneness,
this dreadful connectedness
to all that exists?
I pinch myself to determine
that I am indeed awake,
and drown in the screams
of a trillion creatures.

Jez Green
Manchester

When we get Old

We are planning what we will do when we get old.
Only we won't.
Seeing other people's babies
makes me anxious.
I flit the dial between despair and denial
like a man who can't decide where to sit on the bus.

We are all going to die,
says the woman in BBC Estuary English.
It sounds so melodramatic,
but I can hear extinction
humming in the fridge and the radiator
as another tonne of ice
slips into the Arctic sea.

Take each day at a time, they say,
live in the moment.
As if that will help us survive.
As if all the saved moments make each day
appear, money-back guaranteed.
People say I take it all too seriously,
that I should lighten up,
as we shuffle heads down towards ecocide.

I know I am part of the problem.
Cling onto tiny actions like lifeboats
while others fight with bloody hands for our world.
But I want to do something,
more out of selfishness than anything
because I love this life, and this world and the people in it.
But sometimes the something gets so big
I struggle to see what it is.

Still, I send birthday cards
and joke about our care home
as if it's all going to carry on.
As if life is an all-you-can-eat option,
and we've already paid.

Amy Deakin
London

For Planes to Fly

For planes to fly
they needed land.
They needed meadows,
becks and woods.
They needed gardens.
They needed setts and dens
and forms and burrows,
dewponds, farmyards, one old farmhouse
sheltering downwind of flight paths.
Barn owls evicted, a heronry unfledged,
crow colonies uprooted, scattered,
foxes and elusive fallow deer exiled,
migrants of the edgelands.

And where I used to watch the milk-breathed cattle,
hares boxing among whiskered grasses;
where yellow rattle, clover, chamomile,
and thyme beguiled small bees,
whole galaxies of butterflies
and stained-glass, steampunk dragonflies,

a pyroclastic flow of runways, aprons,
hotels, car-parks smelt the landscape now,
long draughts of tarmac crowned with concrete,
set with steel and glass – the sheen of progress.
And when we said we minded –
for the beetles, for the tiny spiders, for palmate newts
and small, brown frogs, discreet, squat toads,

for fish fry in the beck, wild orchids,
a parlous glint in primrose banks –
they frowned, noodled their brows,
their environmental dyslexia profound.

For planes to fly, we need the land.

                                     Lesley Quayle
                                     Dorset

Blind Man's Buff

I see your dead stick tapping on my pavement
and mark the outward signs of the blind man,
white stick, dark glasses, well-trained dog, all three.

I see white chalk-marks on a healthy tree.
Walk through these beechwoods, trees are overturned
with dead-white chalk in huge lumps clogging their roots,
sprawled in dog-mercury, looking at the blue sky;
dismantled trunks, and boughs lopped off to burn,
twigs deep in nettles, giants brought to their knees.

I rub my aching lids. The trees must die,
and men with dogs walk through the bright green shoots
carrying maps, mark down the doomed trees.

                                                Merryn Williams
                                                Oxford

Litter Picking in the Mariana Trench

Victor's Triton submarine,
lowered into Pacific waves,
cuts loose from its mother ship
dives into the abyssal zone

to discover new species –
supergiant amphipods,
spoon worms, a pink snail fish –
for this is where life began

at a mid-ocean ridge,
a tectonic mountain range
with towering hydrothermal
vents billowing black smoke.

At seven miles down,
the sub reaches new depths
when it catches, in its headlights,
sweet wrappers and a plastic bag.

Clare Marsh
Tonbridge
*

the small cabin

imagine being the person who has to deliver the fuel

the smell of the petroleum in your hair on your hands

stopping at a roadside café for a hot tea & a burger

fried on a griddle & served in a freshly toasted bun

the rumble & complain of your engine as it strains

to pull that articulated weight that fat & silver cigar

accelerating up every three-lane blacktop's slow-lane

gently leaning into every corner but unable to u-turn

as there's never room to driving through the driving

rains & hails & sometime snows nothing to stop you

getting to where you've been paid to go the sun hot &

reflected in your right wing infusing the atmosphere

making the small cab you sit in uncomfortably warm

Jack Houston
Hackney

This is the Last Time I will be a Bird

This is the last time I will be a bird.

My belly swoops full of plastic and ocean,
these wings become arms as I fall
through insignificant thermals.

Here in my new thin fingers I grasp
land like a rug. Where feathers may have
brushed their combs so lightly to leave barely a mark,
I can only sculpt my presence into the earth.

I bleed here.
The sea washes me clean, salts my wounds.
The sun drys me out so I scab over.

Aloneness becomes loneliness then isolation.

With a fist full of shells and a storm coming in,
coral becomes blood,
or blood becomes coral,
brittle with life's leaving.

The contents of my stomach
live beyond me, making ever smaller grains
of colourful sand.

Susannah Violette
Neestahl, Lower Saxony

little funerals

I've only recently started making little graves out of things

just yesterday I made a grave out of an abandoned 'save the bees' cardboard sign, made another out of my rather abused, dog-eared, 101 places to visit before you die guidebook and one for my ice cream, undulating into a placid puddle before I've even licked it once

I didn't want to but I did anyway, make one out of an overturned recycling bin, clogged with snotty tissue, sticky sweet rind, and plastic wrapper, plastic bottle, plastic fork, spoon, straw, knife, *knife*, hand me a knife for a post-mortem after a plastic apocalypse — *too many graves for that already* — but oh let me make one out of a whale's bloating belly, yawning mouth, an open wound, inhaling — mouth

mouth, yes, I'll make a grave for mouths, for the half-lies, half-truths, zero promises, no guarantees, time's up *goodbye*, all climate emergency this, climate emergency that, but sudden dementia — and *oh, sorry what did I say again* — climate science is a hoax, hush now sleep a little better tonight I'll make a grave for our playground politicians —

but also for that teacher on his tenth day of a hunger strike, for the raw red wrists of a pregnant mom chained to the parliament building, for that girl, yes her with the braids, sat on her own, jaw set, back straight, on the school steps even after the bell has gone for the third time

I'll make a grave for all of them, all their hoarse voiceless voices, bless them, *truly* bless them, and amen, may the dead finally speak louder than the living for don't worry I didn't forget to make one for our coral reefs, scrapped, skeletoned, spat out like a slimy chicken bone, or one for the black lungs of our rainforests, charred, but still smoking, still on fire, like our planet, like our damn house is and always has been, call 911 and no one will answer, they choose not to

a million miles away from me a father wakes up to his new-born, wheezing like a hissing kettle, and he checks for pulse I leave the innocent for last, for I also made a grave out of two cotton baby shoes bobbing above muddy water, maybe another one for the mom cradling them to her chest like a talisman, one for the parched panting tongue of a carcass child, white-swollen ribs burgeoning into skin, another for the teetering feet of the last polar bear lumbering in search of a new place they can call home

home, now this one gets a special grave, this one I bury with my two unborn children, fold and tuck them in, throw in a bit of *future* with them, for safekeeping, then press their starry eyes shut and with their tiny fists

smother them dead

Ashley Chew
Singapore / Edinburgh

How I Mistook the World for a Cow

It's true. I thought the world was a large cow.
I used to drain her bulgy dugs most days.
But blame the world for this. The world allowed
me to believe she was a cow, the way
she'd moo on cue and squirt out milk to drink.
When I tied her up too tightly, somehow
she only twitched her tail and let me think
she *loved* to roam around and play the cow.
I mean, she never tried to get away.
She never bellowed to be fed or sheltered,
just stood, slow-blinking, at what came her way,
unmoved by belching trucks and cars that pelted
past. Oh well. So, the world is not a cow.
How will I fill my empty stomach now?

Cathy Dreyer
Oxfordshire

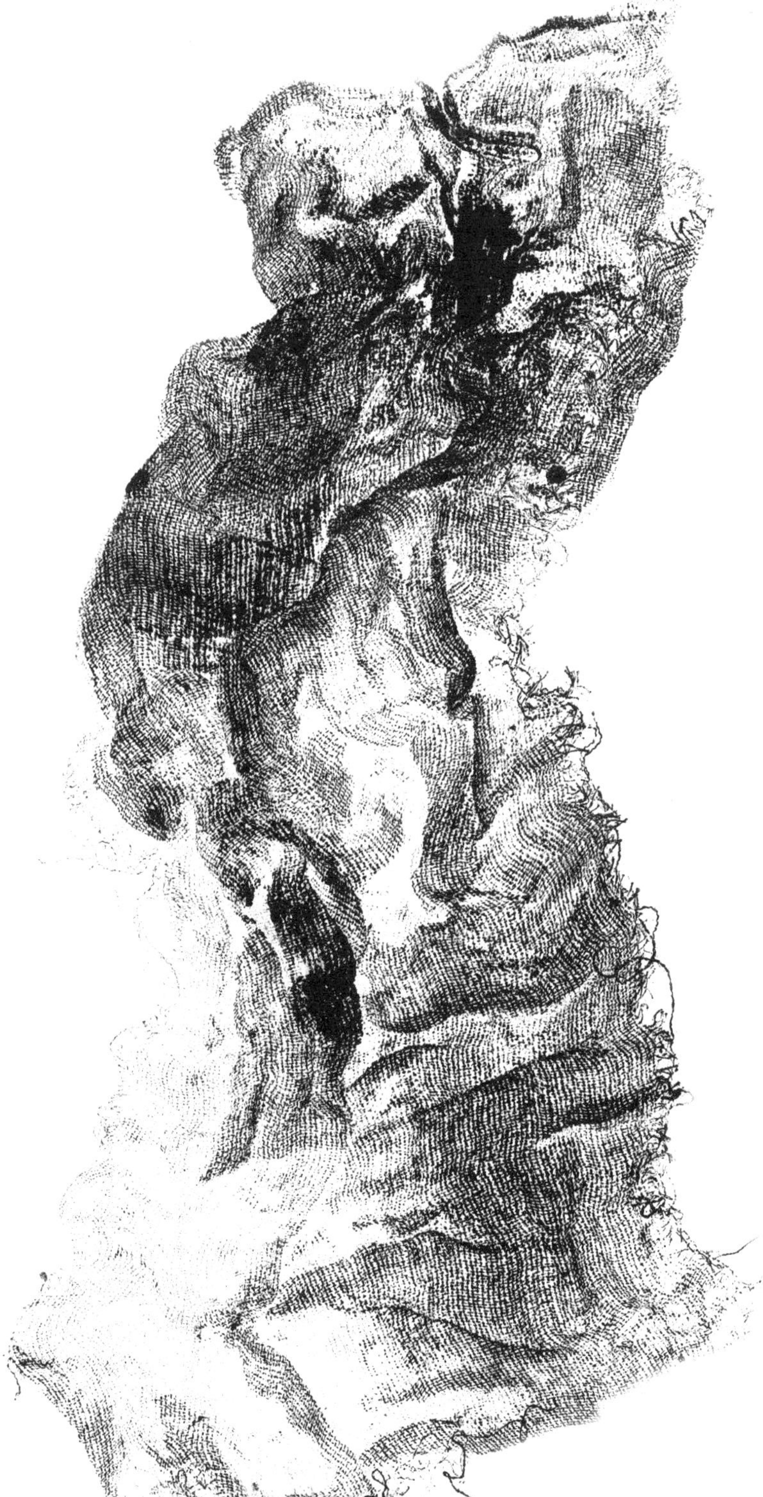

STRUCTURAL

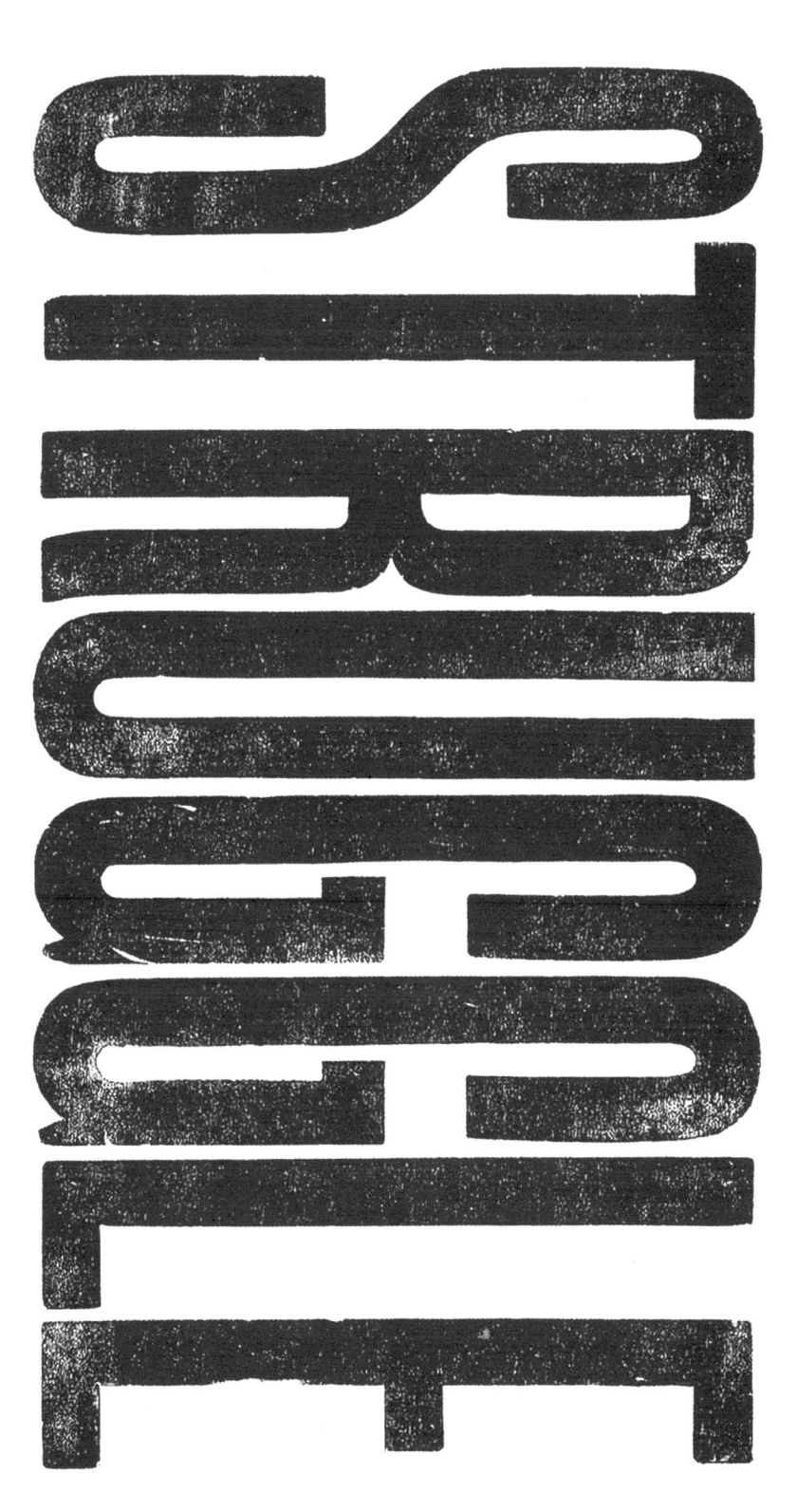

Youth Strike

I missed some school. I told my friends,
But what can I do to make amends?
What scientist say gives me great fear;
What grownup are going to do is what I want to hear.
To reverse climate change what can I do?
I still have to ask teachers when I need the loo!

Aaron Lomas
Oxford

The Coral's Grief

Do not concede the gentle coral's grief
Attend the lessons nature has to teach
Rage, rage against the dying of the reef

Great oceans flow too warm and not too brief
So polyps spit out algae, turn to bleach
And gentle corals fade to death-white grief

The gas-cooked greenhouse climate is the thief
Of life from barriers great and praised in speech
Rage, rage against the dying of the reef

The profit-maker and the finance chief
Will calculate the budget will not reach
Enough to stop the gentle coral's grief

Gaze down glass-bottomed boats in disbelief
At global treaties vetoed, sunk and breached
Rage, rage against the dying of the reef

The tide is running out to bring relief
To fight to save the life beyond the beach
Do not concede the gentle coral's grief
Rage, rage against the dying of the reef

Janine Booth
Lewes

Why should we Care?

Love doesn't form you, experience does. You know how smart you are.
But sometimes your memory fails you - like that time you said 'never again' and still, you did?
Remember how afraid you were (we all are) in the beginning –
we want the freedom to go out and enjoy the space and when it's dark, to come back inside.
Life is a template that keeps renewing itself, but we are good at forgetting.
That wheel that served us so well in the first place (we moved mountains, didn't we?),
re-invented, never turns out to be quite so perfectly round.

I want to tell you a story about geese – The way they fly, arrow shaped,
and how they get to where they are going. There is always the one out front,
but to lead is tiring and to only ever follow one head is unsustainable.
So positions change. They have to. Each goose lifts the air with her wings
to make the path a little more navigable for the one behind, and so on. You get the drift.
And when one goose falls, two others will fall, deliberately, to be with him.
It's not a friend – or even a kindness – thing. It's survival. We stay alive by looking after one another.

Let me tell you another story, about a child pulled from the wreckage after a bomb,
or an earthquake, a fire or a flood, decimates the building and the bed she was asleep in,
together with her baby brother, her mother and father in the room next door,
her cousins and her grandmother sleeping in the room next to that one.
The rescue worker who finds her, carries her blinking into the daylight
where there are lightbulbs and thermoses of hot drink. Cameras and people cheering.
'Now we need to find you a home,' one of the people says, pointing to the rubble.
'I have a home,' says the girl, pointing to her heart, 'I just need a house to put it in.'

Before now, I had never been up close to an iceberg or considered it anything other
than metaphor – some cold sinister presence just under the surface that might,
if we're not careful, break open the hull of the safe ship we are sailing in and sink us all
into oblivion. But now I can see that it is the ice in peril, and we are the danger
standing helplessly by while vast parts of our world calve into the sea
disappearing the fish, the fox, the ox, and the caribou – floating them away
on islands of false hope – the polar bear escaping hunger and guns
and the gush of meltwater flooding its homeland, sinking us all into oblivion.

Some weeks ago I pressed my face next to a block of ice
lost from Greenland's ice sheet, fished from Nuup Kangerlua fjord
and brought to rest at London's Bankside. I held myself awhile
against its dying coldness, looked into its impossibly blue interior,
listened to the drip and crackle of its ever-diminishing tonnes
and ran my tongue across its surface to try to taste its ocean past;
which, I discovered later, at home in bed with the one I love,
tastes the same as the salt we make with our bodies
when we sweat, or bleed, or cry.

Cheryl Moskowitz
London

Glyphosate
to an absentee landowner

Oh don't spray death across my field,
this wiped-out spring insults the soul.
Reckon beyond this year's yield,
not just the profit, but the whole.

We locals have no power to wield,
to save the so-called weeds you kill.
It's money in your bank, I know,
to herbicide our lovely field.

The year's best time, we fear to go
out in the morning into brown
where green should spring and everything
lies on the panned soil, rotting down.

We're unentitled, but we love
clover along the public path.
What ancient moot could have foreseen
this power to lay waste all that's green?

Your hirelings come from far afield
and never walk the land they spray.
Seated high behind a shield
they do the job and drive away.

Airborne poison uncontained
drifts beyond your acreage,
and what you take's not yours to spend,
but common ground along the edge.

Sarah Watkinson
Woodstock
*

play-worn globe
peeling at the poles —
it's a look now or never learn
it's all of our turns to turn
this world around

Helen Buckingham
Wells

The Heat

The land sealed over like cast bronze.
We skitted about like lizards
on a graveyard wall,
threw glances up at the silence
beating its horrible wings
      over our heads.

And nothing from the judges
and nothing from the councillors
and nothing from the architects of tomorrow.
Just this deep daylight blue panopticon
      getting quieter.

In the roar of traffic, silence.
In blaring wide-screen lounges, silence.
In frothing high streets Saturday afternoon, silence.
In the aviary banter of parliament, silence,
      an oven over the earth.

The whole island still as a desert night,
the clink and crack of it turning to glass
      and all the bushes burning.

Jos Smith
Norwich

There are no Words on a Dead Planet

Be yourself. Be one of many.
Shout with the only weapon you have.
Be nicely raging. Be credible.
Be insistent. Be alive. Be floppy
in their arms (it hurts less.) Be equal
to it. Be on TV. This revolution's
in 5G. Be there for your kids. Block
roads, planes, boats, banks.
Be carried to a van in a cage of hands.
The custody-man may let you read
if there's nothing in your book
there shouldn't be. Be *nice* to him,
it's a long night with the light on
and no belt for your waist. Be clear.
Be there for police who can't be;
for your step-dad who never said
anything useful, ever. Be precise.
Be the obvious. The guy next-door
in his addiction, shouts for the huge
carbon footprint of his next fix.
Like a lot of us, really. Be there
for kittens; be there for tigers,
for stinging-nettles, wasps, roses;
for bees and beetles and flying ants;
for elephants; long grasses. Be kind
when the guy in the next cell bangs
on the metal to be let out.
Be crammed-in. Be smiling. Be happy
when they set you free. Be joyful
to return. Be out there. Be out *there.*
Be courageous. Re-set. Go again.
We have the bandwidth. *Shout!*

Ken Evans
Matlock

this stanza ...

this stanza is set in a future
in it our fate is set
a reality where we did nothing
time ticked on but we ignored
all the evidence around us
in the end the earth burnt up
poor stupid us

this stanza is set in a future
in this we go on living
full of everyway recyclers
it's set in a thriving environment
and a cast of happy humans
the critics gave their verdict
as absolutely epic
mixed with human error

this stanza is not set in a future
in fact it's set in the now
where the final scene is not yet written
still have eight years to choose one

Nor Staines Shaw
Oxford

HOPE

Spring 2018 – Rebel Dawn

Six-thirty, Monday morning,
the busy road empties into silence

    for more than twenty seconds.

Long enough to scent
    cut grass on the soccer field,
hedgerow hawthorn blooming.

And the stillness beats out
    life-drum silence, underground, overhead,

where country meadows
    sink into green,
where city streets hold their breath.

In the Arctic, the glacier pauses

    for more than twenty seconds

in the desert, the soldier halts,
    his trigger-finger slow,

the knife feels heavy
    in the hit-man's hand,

the bomb-aimer thinks of his son,
    his attention faltering

    for more than twenty seconds

no bomb falls
> no ice melts
> no heart breaks.

On the planet something wakes.

The hawthorn blooms and blooms
> into silence, into grace.

And, more than this day, something's dawning,
> out of nature, out of pace,

an ardent, fresh, rebellious race.

Six-thirty, Monday morning.

Joy Kenward
Weston-super-Mare
*

Moving Images

The soundtrack to the film *2050 The Last Generation Speaks*
has been lost.

It vanished with the artworks of the London and New York galleries
and the burnt-out desert-stops of Angkor Watt and Great Zimbabwe.

Like a cry in a storm it came and went
then spread upwards on light wings,
to hover as a message in the sky
warning of Easter Island and Mohenjo-daro.

It had served its purpose.

For us, it was a film in the head, rerun so often
that the images had become Hand-of-God slides
streamed at all hours onto walls and ceilings.

Seen by the elderly it flashed up 1930s photos
of empty shelves and queues at the bank.

Reviewed by the screen buffs it was a switch
from unreal horror to a death in the living room,
while for the dedicated gamers it was WW3
fought for real in hand-to-mouth emergencies.

But then came the remake. Pieced together
from memory and scraps left over,
it told the history of die-ins, vigils,
lock-ons and ceremonial occupations,
and the worldwide Children's March
from capital to capital speaking truth to power
that brought things to a stop.

And now, in playback, we're caught on camera
blocking traffic, flesh against metal, protecting the wild,
then talking satyagraha
and being the change you want to see,
as we walk with Siddhartha in the woods
while hearing the sound of
trees in the wind, rain on grass, animals feeding.

                                                                Leslie Tate
                                                                Berkhamsted

I saw our world upon the ground,
Crawling on its belly.
I watched it for a little while,
Then turned back to my telly.

But in my soul I heard its plea,
Groaning and expiring.
I knew it could be so much more;
There is no Planet B.

I joined my local group XR,
Small and young and hopeful.
We disrupted everyone's days,
And turned our planet beautiful.

Valerie Leppard
Poole

farigoule

earth the singular essence of what we are
fluted corollas, eye's wide corridors, the Luberon hills
skeins of mist forming and evaporating
taking a few minutes to stand and feel in space
how it all spreads, how it always has, with
multicellular vista and that we are only one of its orbits;
appearing and vanishing-points are everywhere
around and within, and if we revel at ions
and invisible atoms so the hawk's eye and weasel's
and all the micro-molecular intersections
are matrix for our celebration and fear
of losing this; and it doesn't stop there
these quiet guests in offices, these visitors
in the house we call human are only infinitesimally
part of it, that is I and you and them recurring
and realize this is in our hands; we have this era
outdistanced our capacity to care it seems for wonder

the scent and pink flowers of wild-growing thyme
or farigoule which gives this place its name
and the green-silver leaves so delicate yet resistant
to the mistral's blowing over terraces and olive bark,
flourish in scattered places, past potsherds and bone
and, hopefully, long after this endangered outline
has weathered that which we have made a storm,
will come to add savour for those who wander through,
if I and we and you can re-discover a way to go

David Annwn
Yorkshire

Anthropocaust

Cassandra never smiled. Devastation,
though future for others, was her constant
reality. En route to the death camps
people were still together, had luggage,
and could try to look forward. Even after
being stripped and selected, there may
have been distractions, conversation;
but oven fodder did not smile.

Our delusions are the disgraceful
mystery of these times: wars, festivals,
work, murders, pets, vaccines, shopping,
charities, music, and so unbelievably on.
Meanwhile the human vehicle is colliding
with signposts, shedding fenders against walls;
its tyres are fiery Catherine wheels. But we,
the passengers, turn inwards to shut out
the calamity we have long invited
and do not intend to overcome.

The climate campaigners offer comfort
to some who despair, suggesting they turn
away for respite with the fantasies of life.
It comes naturally from those who must
self-deceive about the power of governments.
But no wonder that so many others
are still preparing for fulfilment,
falling in love, making homes,
and sharing their enthusiasms
for every variety of sport.

Rip Bulkeley
Oxford
*

And what if all we do bears fruit?
What if our children's children live?

What if extinctions,
not species, disappear?

Why be certain we have lost?

As we crest the wave
of collective action's
unknown outcomes,

let the risk of loving
be our singular hope, and

uncertainty be our friend.

> **Justin Kenrick**
> Edinburgh

The Good Ancestor

Every day I walk a hundred years
to the hill where my great great granddaughter sits.
I carry words of blessing
and reach to touch her back.

But feeling me near she turns
sad eyed and heavy with grief.
'What was it like?' she asks
'when the great whales swam
when the birds sang you awake
when the rains came soft
and the soil smelled sweet underfoot?'
And the blessings
catch in my throat.

On darker days she turns,
her famished face charred and eyes,
sunk in their bony orbits,
burn with curses.
And the blessings
froth at my mouth
with the poisonous
spume of betrayal.

On the darkest of all days
I walk the hundred years
and find no one there.

Let today be the bright day.
Let today be the bright day
I lay my hand upon her back
and, feeling me there, she turns
and blesses me, saying
'Your love was fierce enough,
sweet ancestor, your love
was fierce enough'.

Daverick Leggett
Dartington

still life
for a critically endangered stick insect

feather-light

the weight
of an unblemished
soul

so slight a life
you think Atlas would
never know the difference

but her death
would hang most
heavy

being the lone
parenthesis
between her kind
and extinction

curled in your hand

she has folded
six legs with the dignity

of the vanquished

and if her eggs
should fail

the meek
inherit nothing

but you
go on all night

with your pipette

and it is worth
clutching at straws

to feel her tiny
origami form reviving

yes
when it's time
to balance everything

and reckon
all we've taken

lay her in the
other scale

she's like a thought
you're holding

Emma Jones
Abingdon

Notes and Acknowledgements

Golden Child The undulate ray, once common in British waters, has now been classified as endangered by the International Union for Conservation of Nature (IUCN). Last seen in 1989, the golden toad has now been classified as extinct by the IUCN.

Ocean Compost Phytoplankton take in 400,000 tons of CO_2. Their activity accounts for 50% of the world's photosynthesis and generates about 20% of the oxygen on the planet.

Dead Ice: the End of a Glacier was first published online in *Allegro*.

California Screaming The song California Dreamin' was written by Michelle and John Phillips in 1963 and made famous by The Mamas and the Papas.

Mountain Language appears in the author's collection *Skookum Jim and The Klondike Gold Rush* (Indigo Dreams, 2020). A revised draft was first published online in Reuben Woolley's *I Am Not A Silent Poet* (WordPress) in 2015.

Rainbows Behaving Badly 'Even the rainbow has a body' – D.H.Lawrence.

Gorge was originally published in *Bind* and in *The Isambards' Botanic Verses* (University of Bristol), both in 2019. Fossil shells and corals indicate that the limestone of the Avon Gorge formed in shallow tropical seas in the Carboniferous, 350 million years ago. In 1470, workmen rebuilding St Stephen's church in Bristol found the remains of a boat with a mast and a striped sail, dating from before the River Frome was diverted in 1248.

Inundation was published in the collection *Wherever We Live Now* (Red Squirrel Press, 2011).

Harbinger was originally published in *South Bank Poetry* and then in the author's collection *Elastic Man* (Indigo Dreams, 2018). 2015 saw the mildest UK December on record – so far.

Let there Still be Birds was published in *Dream Catcher # 38* in 2018.

Icebergs at Tate Modern In December 2018 artist Olafur Eliasson and geologist Minik Rosing installed *Ice Watch*, a group of twenty-four blocks of ice, in front of Tate Modern. The ice-blocks were lifted from the Nuup Kangerlua fjord in Greenland after becoming detached from the ice sheet.

How to Bake a Planet was the title poem in the author's recent collection (Salmon Poetry, 2016).

The Naming of Storms Henry Reed's poem *Lessons of the War: I: Naming of Parts* was first published in *New Statesman and Nation* in 1942.

An earlier version of **Connected** was published online at WordPress: https://middleofeverywhere.wordpress.com/2017/10/23/moment/

For Planes to Fly was first published in Roselle Angwin's blog *Qualia*.

Blind Man's Buff was first published in the author's collection *The Latin Master's Story* (Rockingham Press, 2000).

Litter Picking in the Mariana Trench In May 2019 Victor Vescovo, who created, sponsors, and pilots the Five Deeps Expedition, found plastic in the 10,900m Challenger Deep.

the small cabin was previously published in the anthology *Welling Up* (Palewell Press) and in *Brittle Star* magazine.

The Coral's Grief was published in *Disaffected Middle-aged Women: poems by Janine Booth* (Roundhead, 2018).

Why should we Care? was commissioned by Helen Yeomans for her *One World* choral project, and performed at various London venues in April/May 2019. It was also performed on XR stages in London, at Westminster Bridge in April, and at Horseguards Parade in October 2019.

Glyphosate A version of this poem was published online in *I Am Not A Silent Poet*, edited by Reuben Woolley, in May 2016. Glyphosate is a widely-used organophosphorus weedkiller routinely sprayed across fields to destroy winter cover crops, so preparing the soil for drilling seed in spring. It is not specific to particular weeds, but kills all wild plants.

'play-worn globe...' appeared in the anthology *Earth: Our Common Ground* (Skylark Publishing, 2017).

The Heat was previously published in *Subterranea* (Arc Publications, 2015).

Spring 2018 – Rebel Dawn Extinction Rebellion was established in May, 2018. Knowing nothing about it, but at around the same time, the poet, on her way to work, found herself walking into what felt like an energetic silence – an inexplicable moment of conception and hope.

Moving Images also appears in *We are a many-bodied singing thing* (RSPB 2020).

Anthropocaust has been published on the Culture Matters website and accepted for the *Oxford Magazine*.

'And what if everything ...' was first published in *Equinox* on 21 March 2019.

About the Poets

Shanta Acharya is the author of twelve books. Her latest publications are *Imagine: New and Selected Poems* (HarperCollins, 2017) and *What Survives is the Singing* (Indigo Dreams, 2020).

Jim Aitken is a poet and dramatist who also tutors in Scottish Cultural Studies. He is currently editing an anthology of radical Scottish poetry for the website Culture Matters.

David Annwn's most recent collection is *Red Bank* (2018). His poems have been filmed by Howard Munson and made into calligraphy books by Thomas Ingmire.

John Barnie has published 27 collections of poetry, memoirs and essays, including *Fire Drill: Notes on the Twenty-first Century* (UWP). A new book of poems, *Sunglasses*, was published by Cinnamon Press in 2020.

Michael Bartholomew-Biggs is poetry editor of the online magazine *London Grip*. His latest collection, *Poems in the Case*, embeds poems in a detective story.

Jordan Biddulph has watched his city (Stoke-on-Trent) begin decaying as it is continually abandoned. He tends to write about industry versus nature and society.

Julian Bishop is a former television journalist who runs a contemporary poetry workshop in Enfield. He was shortlisted for the Bridport Poetry Prize in 2017, and runner-up in the 2018 International Ginkgo Prize for Eco Poetry.

Janine Booth is a ranting, rhyming, revolting poet who lives in Lewes, and tours Britain and beyond bringing poetry to politics and politics to poetry.

Sarah J Bryson is a nurse and a keen photographer. She is interested in words, words for well being, people, nature, and the connections between these aspects of her life.

Helen Buckingham's short form poetry collections include: *water on the moon* (Original Plus Press, 2010) and *sanguinella* (Red Moon Press, 2017), each of which was shortlisted for a THF Touchstone Award .

Rip Bulkeley founded Oxford's Back Room Poets in 1999 and recently edited *Poems for Grenfell Tower* (Onslaught Press, 2018). Like his namesake Van Winkle he got woke far too late but is now catching up.

Rachel Burns was runner-up in the BBC Poetry Proms 2019 competition, and her poetry is widely published in literary magazines. Her poetry pamphlet *a girl in a blue dress* is published by Vane Women Press.

Lesley Burt has been writing poetry for about twenty years. After retiring from social work education, she completed an MA in Creative Writing from Lancaster University.

Miki Byrne is disabled. She has had three poetry collections published and over 500 poems in magazines and anthologies, and has read on radio and TV.

Ashley Chew is just a hopeful university student who wants to be on the right side of history and tell her children cool stories of how the world did not collapse. She is currently doing an undergraduate degree in English Literature at the University of Edinburgh.

Rachael Clyne is passionate about eco-issues and reads at eco-events. Her collection *Singing at the Bone Tree* (Indigo Dreams) looks at our relationship with nature, and her pamphlet *Girl Golem* (4Word Press) at migrant backgrounds.

Alison Cohen strives to marry urban life with getting her hands in the soil, breathing fresh air, and staying in touch with the seasons.

Amy Deakin is a south-west London performance poet and writer. Her debut collection is *Morden and Other Tourist Destinations* (William Cornelius Harris Publishing, 2017).

Jenny Donnison is a PhD student at Sheffield University researching the representation of animals in contemporary poetry. Her poems have appeared in *Route 57, The Sheffield Anthology, Zoomorphic, Riggwelter Press* and elsewhere.

Cathy Dreyer is a poet and critic who lives in Oxfordshire.

Philip Dunkerley has always been passionate about nature. He runs the Stamford Poetry Stanza and performs at open mic events whenever he gets the chance. His poems have appeared in a range of magazines, webzines and anthologies.

Ken Evans won the Kent & Sussex Poetry Competition and published his first collection, *True Forensics*, in 2018. His poems have appeared in *Magma, Under the Radar, Envoi,* and *The High Window*.

Rebecca Gethin has been a Hawthornden Fellow and a Poetry School tutor. *Messages* was a winner in the first Coast to Coast pamphlet competition. *Vanishings* (Palewell) and *Fathom* (Marble) are forthcoming.

Jez Green works for a local charity, tackling poverty, inequality, and homelessness. He is part of the creative collective, Creature. He loves second-hand books and vinyl, and walking in the woods.

Philip Gross has published some twenty collections of poetry and won major prizes, but sees most hope in his collaborations with writers, scientists, musicians and visual artists of all kinds.

Annest Gwilym is the editor of the webzine *Nine Muses Poetry*. Her first poetry pamphlet was *Surfacing* (Lapwing Poetry, 2018).

Martyn Halsall is a former Guardian staff correspondent whose writing is informed by reporting on post-industrial communities, and later becoming first Poet in Residence at Carlisle Cathedral.

Oz Hardwick is a poet and photographer whose latest collection is *The Lithium Codex* (Hedgehog, 2019). He still believes art can change the world.

Deborah Harvey's poems have been widely published, and broadcast on Radio 4's *Poetry Please*. Her four poetry collections are published by Indigo Dreams. She is co-director of The Leaping Word poetry consultancy.

Jack Houston lives in London with his partner and their young family. His poetry has recently featured in *Blackbox Manifold, Brittle Star, The Lake, Nine Muses* and *Stand*.

Caroline Jackson-Houlston started writing poetry seriously after forty-one years academically criticizing and editing it at Oxford Brookes University. Her work concentrates on the natural world and the threats it faces.

E. E. Jones is a writer, activist and community campaigner living in Oxfordshire. Her poems have appeared in *The Journal, Acumen, Stand, Poetry Ireland Review*, and the *Live Canon 2019 Anthology*.

Justin Kenrick is an anthropologist activist. He supports forest peoples in Africa campaigning to secure their lands, and helped establish the first Scottish urban community right-to-buy. Loves parenting, wild dancing, seasons turning.

Joy Kenward's poems often reflect her affinity with nature, which also drew her to her local XR group. Believing everyone can make positive environmental changes, she has re-wilded her front garden.

Emma Lee's publications include *The Significance of a Dress* (Arachne, 2020). She co-edited *Over Land, Over Sea* (Five Leaves, UK, 2015), is reviews editor for *The Blue Nib,* and blogs on WordPress.com.

Daverick Leggett is active in XR, runs a Climate Camp, teaches Qigong and is, on occasion, a poet. He has published one small volume of poetry called *Wild Land, Burning Hearts*.

Valerie Leppard went vegan aged 62 while living in New Zealand, after seeing the way farmed animals are treated and regarded there.

Aaron Lomas writes: 'I am vegan because I don't want to hurt any animals. I love Lego, climbing, science, crafting, cycling and playing outdoors.' He was six years old when he wrote this poem.

Julie Lumsden has been widely published and a poem called 'Godot' is forthcoming in *The Rialto*. She lives close to the famous Crooked Spire with husband and cats.

Mandy Macdonald is an Australian poet living in Aberdeen and trying to make sense of the 21st century. Her debut pamphlet is *The temperature of blue* (Blue Salt Collective, January 2020).

Gill McEvoy was a Hawthornden Fellow in 2012. She has published two collections and three pamphlets, of which *The First Telling* won the 2015 Michael Marks Award. Her forthcoming collection is *Are you listening?* (Hedgehog Press).

Paul McGrane is the co-founder of Forest Poets, Walthamstow. He won the 2017 Geoff Stevens Memorial Award and his collection *Elastic Man* was then published by Indigo Dreams.

Kathleen McPhilemy was born in Belfast but now lives in Oxford. She has been widely published in magazines and anthologies, and her last book was *The Lion in the Forest* (Katabasis, 2005).

Clare Marsh, an international adoption social worker, was awarded an MA in Creative Writing in 2018 by the University of Kent. Nominated for a Pushcart Prize in 2017, her poetry and prose often have dark themes.

Steven Matthews is an XR supporter and has published two poetry collections: *Skying* (Waterloo Press, 2012) and *On Magnetism* (Two Rivers, 2017).

Cheryl Moskowitz is a poet and educator. She believes passionately in the power of writing to bring communities together and effect change. She has published two poetry collections and one novel.

Pete Mullineaux teaches global issues through poetry and drama. He has four collections published, most recently *How to Bake a Planet* (Salmon) – 'A gem' *Poetry Ireland Review*.

Kate Noakes's most recent collection is *The Filthy Quiet* (Parthian, 2019). She is a trustee for the writer development organization Spread the Word. Her first non-fiction title, *Real Hay on Wye*, is due from Seren this year.

Lesley Quayle is a widely published, prizewinning poet, an editor and a folk/blues singer. Her latest pamphlet, *Black Bicycle*, was published by 4Word.

Elizabeth Rimmer has published three poetry collections with Red Squirrel Press for which she is assistant poetry editor. Her fourth, *Burnedthumb*, will be out in February 2021.

Chrys Salt, MBE, has published nine collections, performed her work UK-wide and internationally and been translated into several languages. She is Artistic Director of BIG LIT: the Stewarty Book Festival.

Elisabeth Sofia Schlief's poems have been published in anthologies, online and in her poetry collection, *Es waren Elstern* (bronbiblios, 2015). She takes part in readings, radio broadcasts and book fairs.

Michael Shann has had three collections of poetry published by the Paekakariki Press: *Euphrasy* (2012), *Walthamstow* (2015) and *To London* (2017). He is a member of the Forest Poets Stanza in Walthamstow.

Eleanor Staines Shaw, known as Nor, divides her time between climate activism, being a GCSE student, developing her writing, and hanging out with her black cat. She is seventeen.

Jos Smith is a poet and lecturer in contemporary literature at UEA with an interest in the relationship between literature and the environmental movement.

Paul Sohar has translated seventeen volumes of poetry and published three of his own, the latest being *In Sun's Shadow* (Ragged Sky Press, 2020), besides numerous magazine publications.

Leslie Tate studied with UEA, writing the modern-love trilogy *Purple, Blue, Violet*, a non-binary memoir *Heaven's Rage*, and a triple-author book about childhood imagination, *The Dream Speaks Back*.

Susan Taylor has eight poetry collections and her pamphlet *The Weather House* was recently published by Indigo Dreams. Her new poetry show, *La Loba, Enchanting the Wolf*, is currently in development.

Susannah Violette is a Pushcart Prize nominee. She has had poems placed or commended in several poetry prizes and appeared in various publications worldwide.

Sarah Watkinson is a plant scientist whose pamphlet *Dung Beetles Navigate by Starlight* won the Cinnamon Pamphlet Prize in 2017. She is the first Writer in Residence at Wytham Woods, an SSSI managed by the University of Oxford.

Jules Whiting has an MA in Creative Writing. Her poems have appeared in *Orbis, South, Envoi, The Interpreters House, The High Window*, and various anthologies.

Kate Williams is a children's poet, with contributions to numerous anthologies. Her 'climate change' poems are readily available online.

Merryn Williams was the founding editor of *The Interpreter's House* magazine. Her most recent publications are *Poems for Jeremy Corbyn* (ed.) and *The Fragile Bridge: New and Selected Poems* (both Shoestring Press).

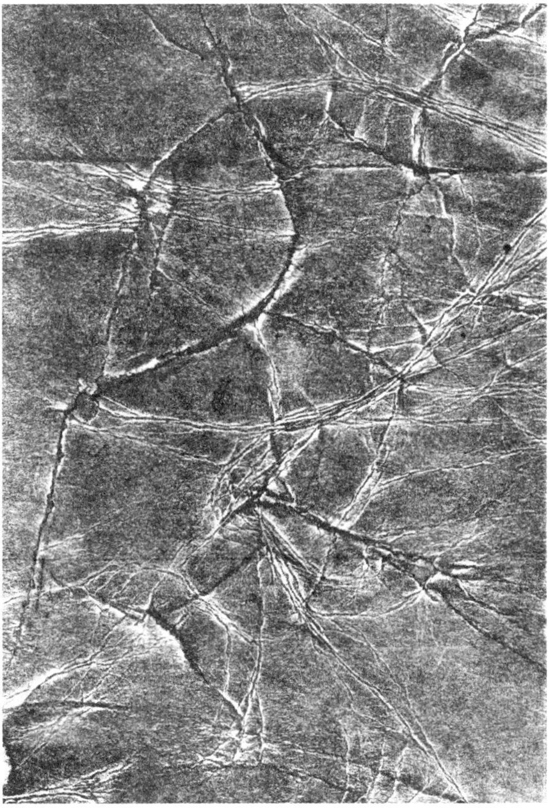